APARTHEID'S GREAT LAND THEFT
The Struggle for the Right to Farm in South Africa

Ernest Harsch

PATHFINDER
NEW YORK LONDON MONTREAL SYDNEY

The articles in this pamphlet first appeared in the December 16 and December 30, 1985, issues of *Intercontinental Press*.

Copyright © 1986 by Pathfinder Press
All rights reserved

ISBN 978-0-87348-487-9
Library of Congress Control Number 2012932016
Manufactured in Canada

First edition, 1986
Seventh printing, 2023

PATHFINDER
www.pathfinderpress.com
Email: pathfinder@pathfinderpress.com

CONTENTS

Introduction   5

Apartheid's great land theft   11

A revolution for Black land rights   49

Appendix:
The land shall be shared among those who work it!   81

Introduction

The land shall be shared among those who work it.

THAT IS THE TITLE of one section of the Freedom Charter, the platform of the South African freedom struggle adopted in 1955 by 3,000 delegates to the Congress of the People.

The Freedom Charter is today the program of the national, democratic revolution to overthrow the apartheid state and tear apart the system of white supremacy in South Africa. That revolution will make possible, for the first time, the forging of a nonracial South African nation-state with full rights for all its inhabitants—the 28 million people who comprise the Black majority, as well as all those among the 5 million whites who are willing to live and work as equals.

The revolutionary democratic movement in South Africa, led by the African National Congress, seeks to abolish all restrictions on the rights of the Black majority to live, work, and travel where they choose. It seeks to guarantee full equality in the job market and to establish full trade union and labor rights. It seeks to replace minority apartheid rule with a democratic republic based on one person, one vote.

The South African revolution, as the ANC explains, has the goal of establishing a single, united, nonracial, and democratic South Africa.

Most people outside South Africa, including opponents of the apartheid system, have little or no access to accurate information about the reality of that society, its class structure, and social relations. The propagandists of apartheid and the big-business press in the United States and other imperialist countries do their best to prevent the truth from becoming known.

This lack of information is particularly gaping with regard to the struggle for equal land rights. Even those individuals who are acquainted with the truth about the apartheid system often know little about the conditions in those sections of the oppressed Black population consigned to miserable poverty in the rural Bantustans or about those being driven in growing numbers from the land they till in the 87 percent of South African territory reserved for "whites only."

This pamphlet is aimed at helping to fill this gap. It is based on two articles that appeared in the December 16 and December 30, 1985, issues of *Intercontinental Press*, a biweekly international newsmagazine published in New York.

By telling the story of the forced dispossession of Africans of the soil and the struggle against this massive land theft, the pamphlet also explains a great deal about the origins of the apartheid system, its current policies, and the revolutionary movement to overthrow it.

The ANC's agrarian program—summed up in the slogan at the beginning of this introduction—flows from the perspective in the Freedom Charter that "South Africa belongs to all who live in it." That program is further explained in the excerpt from a 1969 ANC document reprinted on page 81. The document states that the land to meet the demands of the propertyless or nearly propertyless working people of South Africa will come from expropriating the expropriators, that is, from the large capitalist farmers and landowners who exploit farm labor. The revolution, however, will not take land away from any working farmers; it will divide the land "among the small farmers, peasants, and landless of all races who do not exploit the labor of others."

The centrality of the struggle for access to the land in the South African revolution is a major theme of a report by Jack Barnes adopted by the National Committee of the Socialist Workers Party in August 1985. That report was published under the title "The Coming Revolution in South Africa" in *New International* no. 5, a magazine of Marxist politics and theory. Africans have been dispossessed of their land and forcibly denied the right to farm, the report explains. These are preconditions for the maintenance of the entire structure of racist subjugation under apartheid.

The South African revolution, Barnes says, "is a revolution to conquer the right of the Black majority to own, work, and develop the land from which they have been expelled by the apartheid regime. To win the right of Africans to become free farmers, producing cash crops for an expanding home market. To carry out a real Homestead Act, opening the land to those who want to work it."

Barnes explains that the opponents of the racist apartheid system "get a false picture of South Africa unless we understand the economic and social consequences of [the] forcible denial of Africans' right to own and till the land. If we think of South Africa just in terms of its industry and mines, of what we know about the cities and the white farmers in the countryside, we get a false picture. We see only the South Africa of the white state, of the white minority. We don't see the South African nation-state that has not yet been born. . . .

"Opening the land," the report emphasizes, "is inseparable from resolving the national question" in South Africa. "Neither can be accomplished without the destruction of the apartheid state structure, which blocks the road to development of the South African nation-state." These tasks can only be carried through to completion by a broad revolutionary democratic movement under the leadership of an alliance of the exploited producers—the urban working class, the agricultural wage workers, and the peasants in the Bantustans and in the countryside of "white" South Africa. The toilers from the oppressed Black majority are the vanguard force in this revolutionary alliance.

The information in this pamphlet is a valuable complement to the broader analysis presented in the report, "The Coming Revolution in South Africa." Both are useful tools to help build a powerful movement demanding an end to all U.S. economic, political, military, cultural, and sporting ties with the South African apartheid regime.

Steve Clark
JANUARY 1986

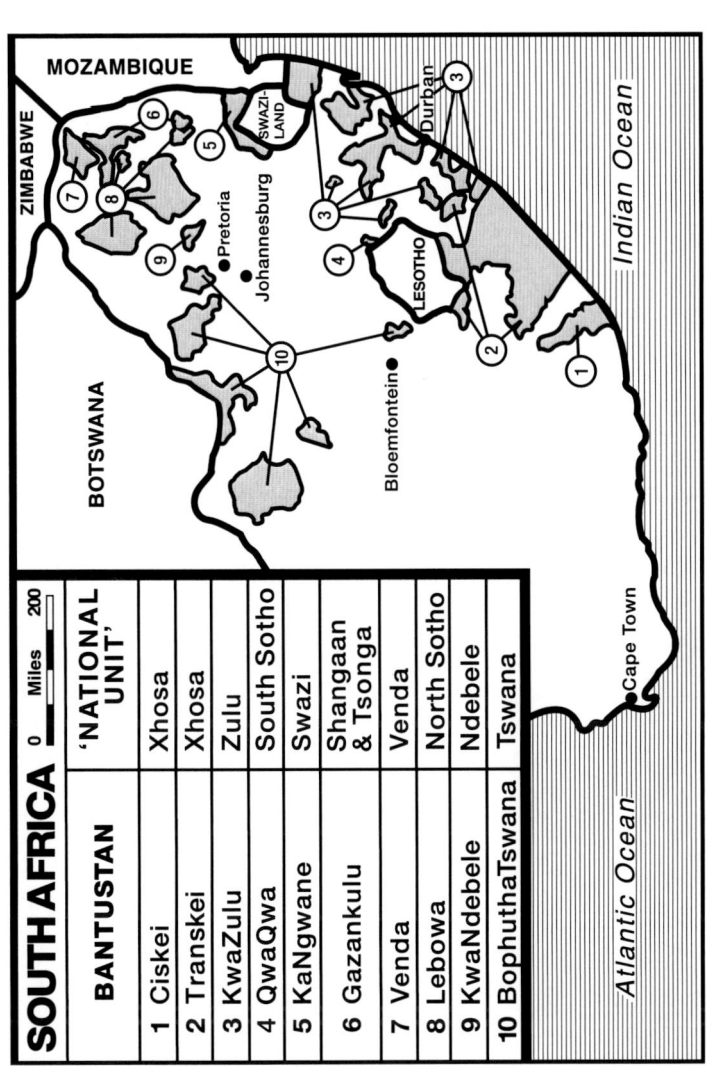

Map of South African Bantustans.

Apartheid's great land theft

> The people are starving. Famine is widespread and will become worse. The people have no land.
>
> CHIEF SABATA DALINDYEBO[1]

FOR CENTURIES LAND has been at the center of the struggle between South Africa's white rulers and its subjugated Black majority.[2] The cry for land has been raised by Blacks time and

---

1. Sabata Dalindyebo was a paramount chief of the Tembu people in the Transkei who opposed the progovernment administration of that Bantustan.

2. The term *Black* is used throughout this pamphlet to refer to the oppressed African, Coloured, and Indian populations in South Africa. The foundation of the apartheid system is the legally established oppression of the 24 million Africans, who are direct descendants of the original inhabitants of what is now South Africa. They have fewer rights than other components of the oppressed Black population. Those of African or mixed national origin whom the apartheid regime has designated as Coloureds number 3 million. There are also nearly 1 million Indians—many of whose ancestors were brought to Africa from the Indian subcontinent as indentured laborers to work on the sugar plantations. Since the 1970s, the term *Black*—earlier used primarily

again, from the earliest wars of African resistance through the current popular upheaval against the apartheid state.

Like the country's vast mineral wealth and the exploitation of its extensive labor power, the bulk of the land is today monopolized by a tiny class of white capitalist families. Originally taken from the indigenous African peoples through colonial conquest, 86.3 percent of South Africa's land is now reserved by law for "whites only." No African has the legal right to own or purchase land in these areas.

For Africans, who comprise some three-quarters of the entire population, a mere 13.7 percent of the land is set aside: the poor, overcrowded, and scattered fragments that make up the ten rural reserves called Bantustans. The other two sectors of the Black population—Coloureds (those of mixed ancestry) and Indians—are also blocked from access to land, except for some small enclaves in the Cape and Natal provinces.

Deprived of their land and cattle and uprooted from their homes, millions of Blacks have been driven to labor for white employers at extremely low wages and under restrictive conditions.

While the white capitalist rulers enjoy the fruits of African land and labor, Blacks are forced to live in the most abject poverty. The urban Black townships are plagued by disease, hunger, and unemployment. Yet conditions in the countryside are even worse. Agricultural laborers must live in wretched shacks and are legally bound to their white employers' farms. In terms of income, literacy, infant mortality, and disease, South Africa's Ban-

---

to refer only to Africans—has increasingly come to be used by Africans, Coloureds, and Indians to identify themselves.

tustans rank with the poorest countries of Africa. They are places of death and despair, where the apartheid authorities seek to dump the unemployed, women, the elderly, the very young—all those not needed to labor for "white" South Africa.

The vast majority of those confined to the Bantustans either have no land or have plots so small that they cannot grow even enough food to subsist; they depend on remittances from the wages of Black migrant workers in their families.

This extreme inequality in land ownership lies at the foundation of the apartheid edifice. It is the precondition for the oppression of the entire African population as a social "estate"—that is, as a sector of the population whose legal and social rights are drastically restricted in comparison to other sectors. This status is codified in law and enforced by the organized force and violence of the apartheid state. The universal oppression of Africans forms the underpinning of the subordinate social and legal status imposed on Coloureds and Indians under the apartheid system, as well.

The denial to the vast majority of Africans of the opportunity to make their livelihood from cattle raising or tilling the soil is the bedrock of apartheid's repressive system of labor control. It is linked to the regime's denial to Blacks of their most basic rights to citizenship. And the political institutions set up to administer the Bantustans are intended to keep Africans divided along language and tribal lines and to separate them from Indians and Coloureds.

Along with the struggle for a democratic republic and for the freedom to forge a modern, unified nation, the fight of Blacks to reconquer the land is an integral part of South Africa's unfolding national, democratic revolution.

Colonial land wars

Before the arrival of the first European colonialists on South Africa's shores in 1652, the various indigenous peoples of the area had access to as much land as they needed.

There were the San hunters and gatherers, the Khoikhoi pastoralists, and the Xhosa, Zulu, Sotho, Tswana, and other peoples who practiced a mixture of settled agriculture and livestock herding. These African societies were poor and their conditions of production were primitive. But they were also relatively egalitarian; they were not plagued by extreme social inequities.

None of these peoples considered the land or other natural resources to be anyone's private property. The land was possessed communally, belonging to the people as a whole. Every family in a village or community had a right to as much land as it could cultivate. Land was divided up and allocated according to local conditions and needs by chiefs or *isibonda* (tribal "headmen"), who functioned as custodians of the land. Unused land reverted to the community for redistribution.

But through a series of wars launched by white colonial authorities and settlers, lasting more than 200 years, these indigenous social systems were shattered. The wars deprived Africans of their very economic foundation and means of livelihood—land and cattle.

The first land war was launched against the Khoikhoi and San in 1658, just six years after the initial white employees of the Dutch East India Company settled near present-day Cape Town. That and subsequent wars took away most of the San hunting grounds and the pasture for the Khoikois' cattle. Whites seized livestock as well.

From the very beginning, the European settlers' land grabbing was intertwined with their desire for cheap, subservient labor. They not only wanted the Africans' land, they also wanted Africans to work the land *for* them. Khoikhoi and San captives were initially reduced to slavery. When they were decimated by a smallpox epidemic, slaves from other African countries and from Dutch colonies in Asia were brought in.

The British takeover of the Cape Colony from the Dutch colonialists at the end of the eighteenth century spurred greater white settlement and led to a rapid expansion of the areas under white control and domination. As the settlers drove east and north in search of additional land and labor, they clashed with more powerful African societies, such as the Xhosas, Sothos, and Zulus. One by one, those peoples were defeated. They lost their land, cattle, and independence and became subject peoples subordinated to the new white *baas* ("master"). New colonial-settler states were established: Natal, which eventually became a direct British colony; and the Transvaal and Orange Free State, which were governed by Boer (Afrikaner) settlers of Dutch origin.

The captured land, which had previously been the communal resource of the African peoples, was transformed into either "crown" land (the property of the British colonial state) or the private property of the European settlers, held through individual freehold tenure.

Some regions remained under direct African occupation and control, however. These areas were administratively designated as African "reserves," the predecessors of today's Bantustans. In them, the land continued to be possessed communally for the most part. Since the settler states were still financially and militarily weak, they preferred to govern these areas indirectly, through tribal chiefs who were induced to collaborate with the colonial authorities. Those chiefs who resisted were deposed. Those who went along with dictates from the colonial masters were rewarded with material privileges. The traditional role of the chiefs and headmen as custodians of the land, reflecting the interests of the African communities themselves, was subverted; more and more of them became privileged functionaries acting on behalf of the white conquerors.

Sharecroppers, peasants, and workers
Before the turn of this century, capitalist relations in agriculture were still very rudimentary in South Africa. Most white farmers did not yet employ Black laborers for cash wages. After the abolition of slavery in 1834, most Black agricultural labor on white-owned farms was organized through a variety of sharecropping and tenant arrangements.

Squatters—as African sharecroppers are known in South Africa—were peasants who lived on white-owned land and farmed part of it with their own seed and implements. In order to use the land, they were compelled to give the white farmer between one-third and one-half of their crops.

Labor tenants were Africans who worked for a white farmer for a specified length of time each year, ranging from three to nine months, in return for being able to live on the farm and cultivate and graze a certain portion of it.

In addition, there was the practice, known among whites as "kaffir farming," under which African peasants directly rented white-owned land.³

These arrangements gave Africans some continued, though restricted, access to land, which provided them with grazing pastures and land they could farm through their own efforts and those of their families. They greatly preferred this to working strictly for cash wages, without any access to land, and resisted the later drives to transform them into wage laborers pure and simple.

Also during the nineteenth century, communities of free African peasants emerged outside the reserves, farming land they had somehow retained or had bought from white farmers. At that time such land purchases were still allowed in the Cape, Natal, and Transvaal, though they were prohibited in the Orange Free State. Benefiting from the expansion of the domestic market, especially with the opening of the gold and diamond mines in the last quarter of the nineteenth century, these African peasants sold part of their produce. Some of them became moderately prosperous, and bought yet more land.

This process caused considerable alarm among the colonial authorities. They feared that the monopoly of the land they had set out to obtain through bloody conquest might be broken by African ownership obtained through the market for land.

The discovery of gold and diamonds in South Africa marked a watershed in the country's economic development. Capital from Western Europe and the United States poured into the country. Capitalist production and market relations expanded rapidly. This further affected agriculture and altered the African peoples' relationship to the land.

---

3. *Kaffir* is a racist term for Africans.

The mining companies developed an acute need for labor in order to exploit South Africa's enormous mineral potential. But initially they could not attract enough Africans willing to work for cash wages—at least for the low wages the mine bosses were offering. Many Africans, moreover, were still able to draw their livelihood from the land and were not yet pulled into the market economy to the extent that they were dependent on cash incomes.

So to create a ready supply of low-wage African labor power for the mines, the British colonial states in the Cape and Natal and the Boer settler states of the Transvaal and Orange Free State enacted new measures, both in the reserves and in the "white" areas, to drive even more Africans off the land. These included antisquatting legislation, new land restrictions, the imposition of taxes, the tightening of pass and vagrancy laws, and the denial of credit and other assistance to African peasants farming for the market.

This offensive was highly successful, from the mine owners' vantage point. The mine labor force on the Witwatersrand gold fields soared from 3,000 workers in 1887 to more than 100,000 African workers in 1899. The families of these workers were forced to stay behind in the impoverished reserves, while the workers themselves labored under restrictive contracts, for specified periods of time. All their movements and residency rights were strictly controlled through repressive measures such as the pass laws, which required them to carry an internal passport on them at all times. When their contracts were over, or they became injured or too old to work, they were dispatched back to the reserves. This was the beginning of South Africa's migrant labor system, which has existed up to the present day.

The completion of the colonial subjugation and destruction of the indigenous African societies and the growth of capitalist relations with the mining boom gave new weight to the efforts of the British colonial authorities to politically consolidate the different states of South Africa under centralized white control. The military defeat of the Transvaal and Orange Free State as independent states during the Anglo-Boer War of 1899–1902 removed a further obstacle to this prospect. Finally, in 1910, the four separate states were unified into a single white-ruled South African state, which at the same time became formally independent of direct British colonial rule.

1913 land act

With the establishment of a single government, the way was also cleared for the white authorities to eliminate the disparities in agricultural and land policy that had previously existed and especially to close off further African access to land.

The 1913 Natives' Land Act—known among Blacks as the "law of dispossession"—opened a broad attack on the few surviving land rights of Africans. It included several main features:

• The act codified in law the white expropriation of the vast bulk of the Africans' land. More than 90 percent of the country was reserved for white ownership and control, including the richest farming and grazing lands, the forests, and all areas with known or potential mineral deposits. No African could own or purchase new land in these parts. Africans were allowed to own land only in those few areas they still effectively occupied and farmed, the areas formally designated as the reserves. At that time these comprised a mere 7.9 percent of the country.

- The act prohibited squatting and "kaffir farming" on white-owned lands. It sought to transform some of the sharecroppers into labor tenants and others into urban and rural wage workers, while driving the rest (particularly non-employed family members) into the reserves. Thousands of Africans were uprooted from land they had continued to cultivate and were forced to migrate in search of work. But many white farm owners could not afford to lose this source of unpaid African labor or offer wages that were competitive with urban industry. So they ignored these provisions of the land act, and as a result squatting survived in some areas for several more decades.
- By outlawing new African land purchases in "white" South Africa and by moving against the squatting system, the 1913 act struck a mortal blow against the emerging African commercial peasantry. Many of those who had already begun to farm for the market were driven out of business.

The white ruling class did not want Africans as free farmers; it wanted them as unfree laborers.

'Land must be in the hands of the white race'
This desire to push ever more Africans onto the labor market was a key motivation behind the land act. As the president of the Chamber of Mines, the mining company association, commented the year before the act was adopted, "What is wanted is surely a policy that would establish once and for all that outside special reserves, the ownership of the land must be in the hands of the white race, and that the surplus of young [African] men, instead of squatting on the land in idleness and spreading out over unlimited areas, must earn their living by working for a wage. . . ."

Child labor in potato fields, Transvaal.

This goal was also recognized at the time by the newly formed African National Congress (ANC). Strongly opposing the land act, the ANC commented in 1916 that the act's aim was "to reduce by gradual process and artificial means the Bantu [African] people as a race to a status of permanent labourers or subordinates for all purposes and for all times with little or no freedom to sell their labour by bargaining on even terms with employers on open markets . . . [and] to limit all opportunities for their economic improvement and independence."

The employers at the same time did not want all Africans to be completely cut off from the land. The survival of the reserves played an important role in buttressing the migrant labor system—the minimal food cultivation in the reserves by the families of migrant workers made it possible for the employers to pay even lower wages. Since these family members were living in the reserves, not in the cities, the ruling class was relieved of the need to confront a large settled urban working class that would demand more and better housing, education, and social services.

But this ability of the reserves to subsidize the incomes of migrant workers was restricted by their extremely small size. Because of overcrowding, more and more Africans became landless, providing a greater spur toward African urbanization and toward the struggle for higher wages in the urban centers. The government's 1932 Native Economic Commission raised an alarm over the deterioration of economic and social conditions in the reserves.

As a result the Native Trust and Land Act of 1936 revised the land allocation provisions of the 1913 act. Promising to "alleviate pressure" in the African reserves, the government undertook to purchase additional land from whites and turn it over to African occupation. The legal limit on African-owned land was thus raised to today's proportion of 13.7 percent of the total land area (although not all of this has yet been incorporated into the Bantustans).

Much of this additional land, however, was already under de facto African occupation, so the 1936 act did little in practice to ease the overcrowding in the reserves. But it did extend the prohibition on rural African squatting to the Cape and lengthened the amount of time that labor tenants were obliged to work for white farmers.

These two acts, of 1913 and 1936, provided the framework in which South African agriculture subsequently evolved. They laid the legal basis for the extreme social inequalities that mark South Africa's countryside today.

White farmers privileged

When the South African government talks about "farming," it has in mind only the white-owned commercial farms.

As of 1983, there were 70,000 such farms in the country, the vast majority of them owned by capitalist farmers or farming enterprises—that is, farms on which the great bulk of the output is produced by wage-labor. These farms account for more than 90 percent of the farm produce sold on the market.

Agriculture's share of the total South African economy has continued to decline. This is because of the growth of mining and manufacturing industries, as well as the racial restrictions that fetter farm production under apartheid. In 1981 agriculture accounted for some 7 percent of the gross domestic product and 9 percent of the value of all South African exports. These exports included wool, maize, sugar, tobacco, and livestock.

The white-owned farms in 1978 covered more than 207.5 million acres of land. Of this land area, some 22.2 million acres were devoted to crop cultivation and the rest to pasturage for livestock raising.[4] Meanwhile, the millions of Africans confined to the Bantustans must make do with 5.9 million acres of cropland and 31.6 million acres of grazing lands.

And compared with the small farming plots of *less than* 9.9 acres each in the Bantustans, these white-owned farms are enormous, averaging more than 2,470 acres.

Actually, the size of white-owned farms in South Africa varies greatly, with most ranging from just under 494 acres to 4,940 acres or more. Some exceed 37,000 acres. The South African government estimates that the top 25 percent of farmers and farm enterprises earn nearly 75 percent of the total net farm income.

The trend over the years has been toward greater concentration of farm ownership. While there are still white working farmers, many of those of previous decades have been pushed off the land in favor of large-scale capitalist farming or have themselves become capitalist farmers. Between 1964 and 1982, the number of white-owned farms fell by nearly a third, from 101,000 to 71,000.

As part of this process, industrial and mining capital is becoming increasingly involved in agriculture. In 1980, for example, the giant Anglo American Corp., the largest mining conglomerate, purchased controlling interest in a third

---

4. Since much of South Africa is arid, and rainfall is erratic, less than 12 percent—some 34.5 million acres—of the entire country is suitable for dryland farming. Of the proportion of this land in "white" South Africa, about three-quarters is actually under cultivation.

of the country's sugar industry, including nearly thirty Natal sugar plantations.

This decline in the number of white farmers will continue under the present system. Even in normal times, the Ministry of Agriculture deems those white-owned farms that are less than 494 acres to be "subeconomic." But the drought of 1982–84 revealed just how economically precarious many of the white-owned farms are, including many above the 494-acre threshold. A survey conducted in 1984 by the South African Agricultural Union, the main organization of white farmers, found that only 54 percent of all the white-owned farms in the country were still financially viable. These difficulties prompted some rallies and other actions by white farmers in early 1985 to protest the high costs of farm inputs.

Many of these white farmers would already have gone under were it not for the special measures taken by the regime to assist the commercial agricultural sector. According to Pretoria, one of the pillars of its farm policy is to "keep white farmers on the land," a policy that itself benefits the richest capitalist farmers over the less prosperous white farmers. This policy of assistance to white farmers is in sharp contrast to the apartheid regime's policy of keeping Blacks off the land. The government's Land and Agricultural Bank provides loans, often at preferential rates, to white farmers (Blacks are excluded from such loans).

More than three-quarters of the white farmers' output is now marketed through state-supported marketing boards. Through these boards, as well as through tariffs, subsidies, export quotas, and other controls, the government has protected the white farmers to an extent from sharp fluctuations in the prices of the agricultural exports on the world market. In fact, it has generally kept agricultural prices well above competitive levels. In 1981, for example, South African maize farmers received $150 per ton of maize, compared to the world price of $111.

Out of the approximately 2.2 million acres of land that are now irrigated in South Africa, all but 86,450 are in farming areas restricted to whites.

Extensive state assistance has also encouraged increased mechanization. Draft animals have now been almost completely replaced by machinery. As of 1977, there were some 285,000 tractors in operation in South Africa—an *average* of four tractors per farm and representing about two-thirds of all tractors in use on the entire African continent. By 1982, total assets of white-owned farms had reached $37 billion.

Agricultural laborers

Whether their operations are large or small, virtually all white farmers in South Africa employ at least some Black farm laborers.

Because of the increasing mechanization of commercial agriculture, these farmers' needs for Black labor have been declining, particularly over the past decade. While there were nearly 1.4 million African farm workers in 1971–72 their number had fallen to 973,000 by 1980. Besides Africans, there are about 250,000 Coloured farm laborers in the Western Cape region. And there are still some Indian farm workers on Natal's sugar plantations, although most have since sought employment elsewhere. About a quarter of all farm laborers in South Africa are women.

Wages for Black farm workers are extremely low. According to the 1976 agricultural census, the cash wages of all regular farm laborers averaged just 32 rands (at that time about US$40) a *month*. In addition, farm workers generally also receive part of their payment in kind, an amount that varies greatly from region to region. This payment usually involves food rations, but Coloured farm workers are given as part of their wage several portions of wine a day (known in South Africa as the "tot system").

Capitalist farmers often prefer to hire women to work the fields. Because of their oppression as women and the few other job opportunities available to them, women are generally forced to accept wages well below the average. In addition, because of the lack of child-care facilities, these women agricultural workers must often bring their children to the fields with them, providing the farmers with a source of extra unpaid labor.

Black farm workers are often compelled to labor fourteen or fifteen hours a day, six or seven days a week. Farm workers' shacks are frequently constructed of dried mud and thatching, with corrugated iron sheeting for roofs; none have electricity.

Beatings with a *sjambok,* a whip made of ox hide, are fairly routine, even for minor infractions. As one white farmer wrote in a letter to a South African newspaper several decades ago, "If we want the natives to be law-abiding, let us speak to them in the language they understand: the language of the sjambok, administered frequently and with vigour." Such attitudes still prevail in the white farming areas. Virtually every year cases come to light of Black farm laborers who are beaten to death; usually the white farmers involved are let off with fines.

Though conditions and wages are poor, it is not easy for Black farm workers, in particular Africans, to leave. This is primarily due to the apartheid system. The pass laws prevent rural Africans from freely moving to the cities in search of more remunerative employment. And once an African is officially registered as an agricultural laborer, it is extremely difficult to change job classifications. Heavy debts to the farmers sometimes place additional bonds on African agricultural laborers.

Like all Black workers, those on the farms are also fettered by antiunion legislation. No unions of agricultural workers have been able to form in recent years.

'Blackening' of the countryside

Despite the colonial conquests, the land acts, and all the apartheid regime's repressive legislation and policies, millions of South African Blacks continue to cling to and farm whatever land they can. They do so not only in the Bantustans, but *outside* as well. The "white" countryside is not all that white.

Legally, these Africans have no rights to own or rent land outside the Bantustans. But the capitalist apartheid regime has been able to enforce this prohibition only partially and with great difficulty.

Some of these African farming communities are on patches of land that were acquired before the 1913 land act and have survived since then. Other African peasants farm white-owned land in regions where labor tenancy, squatting, and similar arrangements still persist.

In fact, with the increased urbanization of whites and the steady decline in the number of white farmers, the proportion of Blacks in the rural population increased markedly in the decades after World War II. More and more white farms were left to the sole occupancy of Black employees or tenants. By 1954, the department of police was estimating that one-fifth of all white-owned farms in the country were occupied by Africans alone.

This trend caused much worry among government officials and strategists, who understood that increased Black access to land would threaten the overall apartheid system, which compels the vast majority of Africans to be proletarians relying on wages for their livelihood. For example, speaking in 1971 in Viljoenskroon, in the heart of a white farming district in the Orange Free State, President J.J. Fouché noted that the Black-white ratio in the district was 16.5 to 1. Such labor patterns on white-owned farms could not continue, he said, because they were leading to a "verswarting" (blackening) of the countryside.

Just as the regime has been engaged in an ongoing effort through the pass laws and other residency control regulations to expel hundreds of thousands of "illegal" Blacks from the cities, it has waged an open war against this "blackening" of the countryside.

There are two main targets of this war: the abundance of Africans living on white-owned land and the hundreds of African farming communities that still exist in pockets of Black-owned land outside the Bantustans, areas called "Black spots" by the apartheid authorities.

During the 1960s, some 340,000 labor tenants and 656,000 squatters (including members of their families) were expelled from white-owned farms. By the early 1970s, the government was claiming that labor tenancy had been eliminated entirely in the Cape, Transvaal, and Orange Free State and that only 16,350 labor tenants remained in Natal. In 1980, the last legal loopholes allowing labor tenancy in Natal were closed.

In addition, the government has also sought to reduce

the number of Black wage laborers on the white-owned farms by establishing maximum quotas for given districts.

Although the government claimed to have virtually wiped out labor tenancy, in fact it still survives. Because labor tenants do not work for cash wages but in return for permission to cultivate or graze a section of a white-owned farm, some white farmers, particularly those facing financial difficulties, have employed subterfuge to avoid having to expel them. Many labor tenants are simply not officially registered as such and are thus overlooked by the apartheid census takers.

As of 1985, it was estimated that there were still some 500,000 African labor tenants and "superfluous" farm laborers who faced possible eviction.

Even with the expulsion of many Africans to the Bantustans, the "blackening" of the countryside continues. One reason is the proliferation of absentee landlordism. In 1981, in the Koedoesrand area of the northern Transvaal, just 320 out of 470 farms were actually occupied by their white owners; in Ellisras the proportion was only 251 out of 654. The remainder, occupied solely by African employees, were owned by white professionals and businessmen living in the cities. Two years later, Gene Louw, the administrator of the Cape, warned that the number of white inhabitants in the Cape's rural areas was declining "drastically," while there was a rise in the number of Coloureds.

Banishing 'Black spots'

After the labor tenants, the other main target of the government's offensive has been the "Black spots." In general, they are farming areas first bought by better-off African peasant farmers before the 1913 land act was introduced, when such purchases were still legal. The land in them is usually owned on the basis of freehold tenure, either individually or cooperatively. Some "Black spots" are on church-owned lands that were leased to African farmers before such renting out was prohibited.

Despite problems of overcrowding, some of these areas remain relatively prosperous—compared to the abysmal poverty facing the vast majority of rural Africans.

The character of these "Black spots" can be seen from a few examples:

- Mgwali, in the Eastern Cape region, is a community of 5,000 people, originally founded on church land in 1857. It is well-irrigated, with three rivers flowing through it. The residents have practiced contour farming for more than a century, and as a result their land has never known soil erosion, unlike many parts of the Bantustans. About 150 landowners possess title deeds, while others are tenants. Farm laborers evicted from nearby white-owned farms have been able to settle in Mgwali.
- Mogopa, in the Transvaal, was first bought by a chief in 1911 on behalf of the Bakwena tribe. Until it was destroyed in 1984, it had three schools, four churches, and two water pumps, all built by the community itself. Occupying two communally owned farms covering 24,700 acres, the people of Mogopa grew maize, sorghum, sunflowers, and other crops. They had adequate grazing land for their goats, donkeys, and cattle. They also owned a diamond mine on their property, which they leased out to a white miner.
- Mathopestad, also in the Transvaal, is a freehold area of about 3,359 acres inhabited by 3,000 people. The land was first bought in 1912. According to a report in the No-

vember 4, 1984, Johannesburg *Sunday Times,* "It is a thriving agricultural community. . . . The people grow mealies [maize], sorghum, beans, and vegetables, besides owning many herds of cattle. They are modern farmers with tractors and boreholes, and for many years have sold their surplus crops to the co-operative in Koster."

The apartheid authorities see the survival of such African farming communities as a direct challenge to their racist system, in which the only role allotted to Blacks is to work for a white *baas.* They have therefore slated these "Black spots" for elimination and have designated them for white occupation.

During the 1960s, some 97,000 residents of these farming communities were forcibly uprooted from their homes and resettled in the Bantustans. The pace of such forced resettlements picked up even more over the following decade.

According to a 1985 estimate by the Surplus People Project, which campaigns against forced resettlements, there are more than 1.3 million residents of "Black spots" and areas scheduled for Bantustan "consolidation" who still face the threat of eviction. In Natal Province alone, it is estimated that 245,000 residents of 189 "Black spots" confront such a prospect.

Sometimes the removals are carried out with the "agreement" of tribal leaders and with promises of new land in the Bantustans. But only those who already own more than 42.7 acres of land—a tiny minority—are actually entitled to compensatory land in the Bantustans. All owners of smaller plots, as well as tenants, are thus deprived of land through the resettlements, and must get rid of their livestock as well.

Left, South African agrarian system also discriminates against Indians. Right, oppressed Black majority has become increasingly united in struggle against apartheid.

Usually, there is no agreement involved in the resettlements—just physical force. George Rampou, a leader of the expelled Mogopa community, explained: "They did not discuss with us. . . . They just come. They come in the middle of the night, all armed with revolvers. They come and surround your house as though you killed somebody. Then they forced you to leave your house without you knowing why, how you must go. They decide how much to pay you without talking to you about it. But you must accept because they already break your house. . . . They must be great cowards to come and surround people when they are all fast asleep to do these things."

Homelands of misery
According to the apartheid propagandists, it is in the ten "national homelands," as Pretoria calls the Bantustans, that Africans can enjoy their political and social rights, including the right to own land and to farm it. But as the millions of residents of the Bantustans know, this is a fraud.

The Bantustans are so overcrowded, fragmented, and impoverished that they provide little possibility for the development of a modern class of farmers producing commodities for the market. In fact, they do not allow most Bantustan residents to farm at even a *subsistence* level. They must rely on a portion of the wages earned by family members working on farms, in factories, or in other jobs outside the Bantustans.

Because of the regime's concerted drive to force as many Blacks out of "white" South Africa as possible, the total population of the Bantustans increased from 5 million to 11 million between 1960 and 1980 alone, rising to 54 percent of all Africans in the country.

None of the Bantustans are viable economic, social, or geographic units, let alone "independent" states, as Pretoria has proclaimed four of them (Transkei, Ciskei, BophuthaTswana, and Venda). Only the tiny KwaNdebele and QwaQwa Bantustans are composed of single pieces of territory, while KwaZulu consists of forty-four patches of land separated from each other by white farming areas.

In recent years, the government has outlined plans to "consolidate" the Bantustans by reshuffling land among several of them and by substituting land that is now owned by whites for some Bantustan territory. Yet this will not alleviate the basic problems. Some Bantustans will gain more land as a result of the "consolidation," but still within the 13.7 percent limit set by the 1936 Native Land and Trust Act. The most immediate effect of this process will be to inflict yet more suffering on the hundreds of thousands of Africans who will be forced to move. As with the removal of the "Black spots," many residents of the areas affected by "consolidation" will lose their cattle and their access to land.

Another consequence of Bantustan "consolidation" will be to further deepen language and tribal divisions among Africans—a key goal of the entire Bantustan program. In the northern Transvaal, for example, the Sotho and Shangaan peoples had close social ties in the past and often lived in the same areas. But the imposition of artificial borders for the Lebowa and Gazankulu Bantustans—sometimes running through the middle of a single village—have fostered divisions and frictions between them. Shangaans who used to rent tractors for plowing from their Sotho neighbors can no longer do so.

Poverty, unemployment, and artificial land shortages

have exacerbated tribal conflicts even more, at times leading to armed clashes. In late 1985 and early 1986, scores were killed in fighting between Zulus and Pondos in the Umbumbulu region of KwaZulu and between Ndebeles and Pedis in the Moutse district of the Transvaal.

With more and more Africans forced to live in the Bantustans, they have become even more overcrowded than before. Between 1970 and 1980 alone, the population of KaNgwane increased by 204 percent, of KwaNdebele by 415 percent, and of QwaQwa by 515 percent. In 1970 the Ciskei had 172 people per square mile; in 1981 it had 326.3. The Bantustans as a whole are among the most densely populated regions of the entire African continent.

When the 1936 land act was adopted, it set a limit of 9.9 acres of land per family—a tiny amount compared to even the smallest white-owned farms. But for most Bantustan residents, even a plot of that size is impossible to obtain. In the Ciskei, the average size of landholdings in the farming areas is only 2.47 acres, which under the Ciskei's dryland conditions is not enough to feed one person, let alone a family. Just 27,000 of the 375,000 rural Ciskeians have enough land to enable them to also keep cattle. Nearly a third of the Ciskei's people have no land at all. In the Transkei up to a quarter of the population is landless.

In KwaNdebele, some 200,000 people who had been expelled from the white areas were each allotted one-sixteenth of an acre of land, barely enough for a house and a small garden.

A reporter for the Johannesburg *Rand Daily Mail* writes in the June 17, 1982, issue: "What strikes you most forcibly as you drive into KwaNdebele, is the way the land changes. As you leave Groblerdal behind, the silver sprays irrigating rich, white farmlands give way to scrubby bushveld and finally to dry, arid soil whose main produce appears to be thorntrees. Dust hangs in the air in the wake of every passing vehicle. . . . Looking around, you notice the bareness of the tiny residential plots, devoid of crops which could supplement the family income. But cows graze on the common land and there seem to be goats everywhere. But it is the aridity that impresses."

Not surprisingly, poverty, disease, and hunger are virtually universal in the Bantustans. Their per capita domestic products are smaller than those of most independent African states. In the Ciskei, 40 percent of the population is unemployed, and 89 percent of the children suffer from malnutrition. In the Transkei, 95 percent of the population have cash incomes below the official subsistence level. In KwaZulu, one in every three boys is stunted in height and weight due to malnutrition, as is one in every four girls.

Given the overcrowding, the shortage of money for fertilizer and farm tools, the general lack of irrigation, and the fact that many adult males are away working as migrant laborers, overall productivity in the Bantustans is particularly low.

In some Bantustans, agricultural production reaches less than 3 percent of its potential. Properly cultivated, the Transkei alone could produce enough maize for all of South Africa, but today it does not produce enough even for the needs of its own people. Throughout the Bantustans, the average per-acre output for maize and other crops is only about a quarter of that on white-owned farms in the rest of the country.

South Africa's overall problem of soil erosion is even more severe in the Bantustans. Because of the conditions under which Africans are forced to work even to eke out a portion of their livelihoods, they are unable to employ methods such as contour farming and crop rotation, which could help preserve the soil. As early as the 1950s, nearly three-quarters of the land in the Bantustans was suffering from moderate to severe erosion. It has become worse since then.

Bantustan tenure

Formally communal land ownership is still the norm in the Bantustans, applying to 94 percent of those who till the soil. But this is not the same communal system that predominated prior to colonization. It has been greatly distorted by the expropriation of most African lands and by the apartheid regime's transformation of most tribal chiefs into appendages of the government administration.

Land tenure in the Bantustans has become frozen. Because of the extreme overcrowding, land is rarely redivided as it was in the past and instead remains in the effective possession of the same families, handed down through the generations with the eldest son as the sole inheritor. Younger sons can no longer obtain new plots of their own. Despite the fact that women head up some two-thirds of all rural households, they have no land rights whatsoever. Under the distorted version of "tribal law" imposed in the Bantustans by the apartheid authorities, women are deemed to be legal minors, whatever their age, with no rights of their own to hold property, enter into contracts, or even act as guardians of their children—permission of a male relative is required for all these. Widows in the Bantustans may retain the right to cultivate the land of a deceased husband but automatically lose that right if they leave for any period of time (such as to seek wage employment outside the Bantustan). Newcomers to the Bantustans—the hundreds of thousands expelled from "white" South Africa—generally can no longer get any farmland.

Although this communally owned land may not be legally sold or rented, sharecropping is permitted, enabling those with land rights to draw a small amount of benefit from the labor of those who have none. Some chiefs demand portions of crops or cash fees in return for making land allotments and have made formerly voluntary tribal tributes compulsory.

But even this distorted communal system is subordinated, in the final analysis, to Pretoria. Families occupying plots of land must pay yearly rents to the local magistrate's office, making them in practice tenants of the state. The government, moreover, has the power to intervene in the Bantustans whenever it sees fit to deny land occupancy to any African. In the Bantustans that have been declared "independent," these powers were simply passed on to the Bantustan administrations. Yet even in these Bantustans, Pretoria can retain legal control over any lands added to them as a result of the Bantustan "consolidation" program.

A small layer in the Bantustans—about 6 percent—cultivates land on the basis of individual ownership, primarily in the Ciskei, Transkei, and KwaZulu. Like the "Black spots" outside the Bantustans, this land had originally been bought before the 1913 land act. It is mainly on these plots that the few self-supporting peasant farmers in the Bantustans can be found. They have higher yields and some even employ labor. While they may sell a surplus from time to time, most farm primarily to meet their own needs and those of their families, rather than for the market.

'Master farmers'

With the direct support of the Bantustan administrations and the apartheid regime, a tiny layer of commercial farmers has also been created in the Bantustans in recent years. Known as "master farmers" or by other designations, they are often well-paid Bantustan officials or tribal chiefs who have used their positions to buy some land or to acquire personal use of communally owned land.

As early as 1974, BophuthaTswana's minister of agriculture, T.M. Molathlwa, was able to comment, "It is notable that in recent times a new breed of farming entrepreneur has emerged amongst the Tswana people. It is not uncommon to find farmers running herds of several hundred cattle. Stud breeders have also been forthcoming, and also in the field of crop husbandry, men owning tractor units and producing up to 6,000 bags of grain per annum are operating on portions of land leased from other farmers or on vacant government land."

In KwaZulu, prosperous sugarcane growers now hire agricultural laborers from as far away as the Transkei. In the Transkei, Prime Minister Kaiser Matanzima and his brother have acquired several farms without any payment.

Whites have also rented land in the Bantustans to set up relatively lucrative commercial farms. In the Transkei alone, white-run farms covered more than 27,170 acres in 1978.

The emergence of this layer of commercial farmers in the Bantustans flows from a conscious effort by the government and the Bantustan officials to build up a political base of support. These farmers are directly dependent on loans, marketing assistance, irrigation, and other favors from the Bantustan administrations.

In addition to these private ventures, the Bantustan administrations have launched other commercial farming operations of their own, including tea, cotton, coffee, and citrus farms.

Such commercial farms in the Bantustans are being developed on the backs of the masses of Bantustan residents. To acquire the necessary land for them, a new phase of land and cattle expropriations has been opened up within the Bantustans themselves, affecting several million Africans.

New taxes have been imposed on livestock, with those unable to pay facing the prospect of having their cattle confiscated. The KwaZulu Development Corporation has taken over the best available land in the Tugela Valley for cotton, wheat, and tobacco farms, while new irrigation canals will divert the river's waters away from tribal lands now under food cultivation elsewhere in the valley. In the Ciskei, the authorities are pressing for a compulsory land "consolidation" and for the imposition of a tax on those migrant workers who have land in order to force them to give it up. In Lebowa, elderly people are being offered tiny pensions if they turn over their land rights.

Apartheid officials have pressed to transfer the occupation of plots of land from "inefficient to more progressive farmers." One official in Pretoria even favored reducing the number of those with land rights in the Bantustans to one-fifth or even one-tenth their present number. The "excess farmers" are to be resettled in urban townships within the Bantustans.

Through a combination of the mass expulsions from other parts of the country and the land expropriations now under way within the Bantustans, more and more Bantustan residents are without any land. In 1960, just 1.2 percent of the population of the Bantustans lived in urban areas; by 1980 17.1 percent lived in townships with an additional 41.6 percent living in "closer settlements" where virtually no farming or stock raising is possible.

Without land these residents are even more dependent than before on working outside the Bantustans themselves or on the income of family members who do so. A 1983 study of migrant workers in Natal found that 46 percent had no land at all, not even shared land. In the Ciskei, three-quarters of all families now depend on money sent home by migrant workers. Seventy percent of the economically active population in BophuthaTswana must work outside the Bantustan, as must 50 percent of those in the Transkei.

Citing an official of the Lebowa Bantustan in the northern Transvaal, the May 14, 1983, London Guardian reported, "In Lebowa, says [Machupe] Mphahlele, land is scarce. Moreover, land that a few years ago used to grow food has now become too heavily populated. More and more able-bodied people have thus had to go to 'the factories in the south'—Johannesburg, Pretoria, and the Vaal triangle—to earn money. They must support themselves there."

Rather than the developing "national states" projected in government propaganda brochures, the Bantustans are becoming more than ever before mere settlement camps for migrant workers and their families. These Bantustan residents form a massive army of unemployed who are readily available as laborers for use by white industrialists and capitalist farmers.

Fetters on Coloureds and Indians

As oppressed Black peoples, Coloureds and Indians likewise face severe restrictions on land rights.

Many of those categorized as Coloureds can partially trace their descent back to the Khoikhoi people, who once roamed with their herds over most of what is today Cape Province and into parts of Natal. They were the first victims of the white colonialists' land wars, and were driven off the land more thoroughly than other African peoples (nearly 75 percent of all Coloureds live in urban areas today). But here and there, in small rural pockets, some of the descendants of the Khoikhoi live in what are termed "Coloured reserves."[5]

According to a government report issued in 1976, there are twenty-three rural Coloured areas covering some 4.2 million acres of land, with a total population of 58,000. Nearly half live in the more crowded reserves of the southern Cape, which consist of only 2.1 percent of the total land area of the Coloured reserves. As in the Bantustans, many residents of these reserves actually hold jobs outside of them, given the lack of sufficient land for farming. There are some Coloured peasants as well in the Transkei, Ciskei,

---

5. Unlike the Khoikhoi, the San (also called Bushmen) were able to retain no land rights whatsoever. Most of the San survivors were driven out of South Africa into what is today Namibia, but several hundred still live within South Africa's borders, mainly in the northern Cape. What remains of their hunting grounds is considered public property, and they must compete, at an extreme disadvantage, with whites who hunt for sport.

and parts of Natal, generally owning their own plots of land.

Ownership of land in the Coloured reserves rests in the hands of local management committees, which, like the Bantustan administrations, are bodies dominated by political collaborators with the apartheid regime.

The residents of the Coloured reserves comprise only 11.1 percent of the total rural Coloured population, which numbers more than half a million. Most of the remainder are labor tenants, squatters, or agricultural laborers on white-owned farms. They face restrictions, difficulties, and conditions similar to those of Africans, although they do not have to carry passes.

Coloureds have no legal land rights in the "white" countryside. However, a few—those favored by the apartheid regime—may purchase white-owned land if they can obtain permits from the Department of Planning.

Like Coloureds, South Africa's Indian population, which today numbers nearly 1 million, is also largely urban.

Yet the first Indians were brought to South Africa a century ago precisely to work the land—as indentured laborers for white sugarcane farmers in Natal. After their periods of indenture expired, many were allowed to stay on in South Africa. Most became workers. A few became urban merchants and businessmen. Some were able to buy or rent land, or acquired plots to farm as labor tenants or squatters. A handful of Indian sugarcane farmers have become relatively well-off, employing both Indian and African agricultural labor.

But like Coloureds and Africans, these Indian farmers are shackled by numerous discriminatory and segregationist laws. They are confined to designated areas and cannot purchase land elsewhere. They can obtain few loans, since credit is available almost exclusively for whites. And over recent decades they have had some 61,750 acres of farmland taken away from them through apartheid legislation and policies.

A backward agrarian system

The social inequities of South Africa's agrarian system exact a heavy human toll on the majority Black population. Hunger and disease are prevalent, both in the squalid urban townships and especially the Bantustans.

It is estimated that nearly 3 million Blacks under the age of fifteen suffer from malnutrition. Between 35,000 and 50,000 Black children die each year of illnesses related to or aggravated by dietary deficiencies. Overall Black infant mortality is high. In some rural areas, between 30 percent and 50 percent of children die before their fifth birthday.

The severe drought that swept South Africa in 1982–84 threw a particularly sharp spotlight on South Africa's social disparities. Its impact on whites and Blacks was very unequal.

Some white commercial farmers reported crop losses of up to 70 percent and a significant rise in their debts. But most were protected from the worst consequences of the drought by state assistance. They had funds from previous bumper crops to tide them over and had access to whatever irrigation was still available. And South Africa's state marketing boards continued to export maize and other grains.

But the peasants and other residents of the Bantustans, who were already living at the bare edge of survival, lacked such state aid. For them, the drought was an unmitigated disaster.

Production of maize, the main staple food of rural Africans, "totally collapsed" in the Bantustans, according to an official of the white National Maize Producers' Organisation.

The March 14, 1983, Durban *Daily News* reported, "The worsening water shortage in Natal is fast spelling financial ruin and an end to the livelihoods of many black sugarcane farmers. Cane production by African, Indian and Coloured farmers dropped by more than 40 per cent in the past year and unless the government came to the rescue of these victims of the prolonged drought many would have to quit the industry they helped to build up...." Such aid was not forthcoming, however.

A week earlier, the Durban *Sunday Tribune* described the impact of the drought on Black farm workers: "The drought has left thousands of farm labourers battling for survival and without work. In some areas they now live a nomadic life and move from place to place in search of grazing and water for their dying cattle. There is no work for them on the farms and they have no means of making money."

By March 1983, an estimated 800,000 head of cattle had already died from the drought in the Bantustans of Lebowa, Gazankulu, Venda, and KwaZulu.

Around the same time, medical researchers in South Africa were estimating that hunger was killing Black children at a rate of at least one every twenty minutes.

This death and misery was caused only incidentally by the drought. Its real cause was the white supremacist system that has deprived Blacks of their land and destroyed much of their agriculture.

South Africa's white rulers claim they brought "civilization" to the peoples of the region, and introduced "modern" farming systems and methods. But the technically advanced

Pondo rebellion in the Transkei, 1960.

farming methods employed in commercial agriculture only serve to enrich the capitalist farmers, not to improve the lives of the oppressed Black majority.

The agrarian system that was introduced into South Africa is among the most socially backward in the world, a system in which the land has been seized by conquerors and the conquered peoples transformed into a social estate deprived of the most fundamental rights. For the millions of South Africa's indigenous inhabitants, it has brought exploitation, hunger, and disease.

As part of their broader struggle against white racist rule, South Africa's Blacks are fighting to break this agrarian system and to return the land to those who work it. They know that only the overthrow of the apartheid state can unlock the wealth of South Africa's soil for the benefit of all its people.

A revolution for Black land rights

To you, the sons and daughters of the soil, our case is clear.

The white oppressors have stolen our land. They have destroyed our families. They have taken for themselves the best that there is in our rich country and have left us the worst. They have the fruits and the riches. We have the back-breaking toil and the poverty. . . .

Over 300 years ago the white invaders began a ceaseless war of aggression against us, murdered our forefathers, stole our land and enslaved our people.

Today they still rule by force. They murder our people. They still enslave us. . . .

They have declared war on us. We have to fight back!

ANC LEAFLET
1968

FROM THE TIME OF THE FIRST colonial land wars, South African Blacks have fiercely resisted every move by the white authorities to take their land and cattle.

Though Blacks were militarily defeated and deprived of

most of their land rights, their struggle for land has continued. It takes varying forms, from informal occupations of unsupervised white-owned lands through sporadic peasant rebellions in the Bantustans.

The struggle for the land has not lost its force or importance over the decades. It remains a vital issue for all Blacks. The white monopoly over most land is a cornerstone on which the entire apartheid structure rests and without which it could not have been created and maintained.

Despite South Africa's extensive industrialization and the growth in the number of Blacks living in the urban centers, land is still an immediate concern to a significant part of the Black population. Because of the regime's massive forced resettlements, more than half the entire African population live in the Bantustans. Several million more live and work on white-owned farms.

A considerable number of African wage workers still have one foot in the countryside. They are not yet part of the hereditary proletariat, who view themselves as belonging to a distinct, permanent working class with no further perspective of returning to the land.

There are some 2 million African migrant workers—nearly a third of all African workers—who labor for periods of time in the cities and must periodically return to the Bantustans, where their families live. Most migrant workers still have access to some land, even if it is only a minimal amount, and they have repeatedly shown their determination to hang on to it.

In addition, many nonmigrant workers in the cities are only recent arrivals and retain family and other social and cultural ties in the rural areas. Even working-class families that have lived in the cities for several generations are affected by the poverty, hunger, and disease of the Bantustans. These conditions drive down the living standards of all Blacks.

Demands and struggles relating to the land have been part of every major period of mass opposition to white minority rule. The upsurge that has been rocking South Africa for the past year and a half has been no exception.

"The black rebellion, which began in the big city ghettos like Sharpeville, Crossroads and Soweto, then spread to small-town South Africa, has now reached the pastoral backwaters of what the South African government calls tribal 'homelands' [the Bantustans]," prominent South African journalist Allister Sparks noted in the October 26, 1985, *Washington Post*.

The United Democratic Front (UDF), the 2-million-member anti-apartheid coalition that has led many of the protests, has called for the scrapping of the Bantustan system and of all laws restricting Black land rights. Other groups have raised similar demands. Supporters of the African National Congress (ANC)—the political vanguard of the struggle to bring down the apartheid regime—have been popularizing the slogan, "The land shall be shared among those who work it!"

The fight of Blacks today to reconquer the land is a central aspect of South Africa's unfolding national, democratic revolution. The demand for land—for a sweeping and deep-going agrarian reform that will overturn the racist and unjust system of land ownership—is one that elicits an immediate response from millions of Blacks.

Peasant revolts

Many areas of the Bantustans have long traditions of rebellion and opposition to the regime's racist agrarian policies. Between 1940 and 1963, major peasant revolts swept a number of them.

These rebellions were provoked by the regime's drive to push even more Blacks off the land, a drive that accelerated with the coming to power in 1948 of the National Party. The new government institutionalized apartheid as official state policy and drastically extended the repressive and discriminatory measures of earlier regimes.

In the countryside this involved kicking millions of Blacks out of farming areas in order to tighten the whites-only monopoly over land ownership and rental. The expelled Blacks were forced to settle in the Bantustans, making the Bantustans even more overcrowded, reducing the size of farming plots, and increasing the number of landless.

The authorities embarked on a massive land "rehabilitation" program within the Bantustans themselves, taking land away from many of those who were farming at a subsistence or below-subsistence level. Cattle herds—a key source of livelihood for many rural Africans—were reduced, as were grazing fields. The carrying of the notorious passes, which had previously been mandatory only for African males, was extended to women, adding a further source of discontent.

Rural residents especially resented the 1951 Bantu Authorities Act. It further subverted the traditional system of tribal chiefdoms by making the chiefs subject to government appointments and salaries and giving them wide powers to allocate land. Under this act, apartheid collaborators were favored, while those chiefs who resisted government policies were harassed or deposed.

While African peasants were provoked by these policies, they were also inspired by the rise in mass struggles in urban areas, including union drives and political campaigns by the ANC and other groups. The close connections between rural and urban Blacks facilitated the spread of such political influences.

The main rebellions were in:

- Zoutpansberg, part of the Bantustan today known as Venda in the northern Transvaal, where peasants rose up in 1941 against land reduction programs. For several years there was "a state of armed warfare," as officials described it, which was ended only by massive repression.

- Witzieshoek (today called QwaQwa), which was swept in 1950 by mass protests and armed resistance against the authorities' efforts to reduce the number of cattle in the reserve. Many protesters were killed or arrested.

- Marico, now part of BophuthaTswana, which erupted in 1957–58 against the extension of the pass laws to women. Thousands of women refused to carry passes. The regime cracked down hard, imposing heavy fines and arresting many, some of them on charges of being members of the ANC.

- Sekhukhuneland, today part of Lebowa in the Transvaal, where peasants rallied in mass meetings in 1958 to protest the Bantu Authorities Act and the racist education system.

- Tokazi, part of today's KwaZulu Bantustan, where peasants mobilized in 1959 to oppose government land "rehabilitation" schemes. After the rebellion was crushed, a number of participants were sentenced to death.

- Pondoland, part of the Transkei, where the most massive and organized of the peasant rebellions broke out in 1960. The main grievance was the imposition of the "Bantu authorities" system.

Pondo peasants held periodic mass meetings to discuss their demands and plan strategy. Their movement, known as Ikongo (Congress), formed a leadership called Intaba (Mountain). They set up people's courts, both to bring apartheid collaborators to justice and to take on tasks such as allocation of plowing lands. They initiated boycotts against rural traders who opposed the movement. In addition to rejecting the government-appointed chiefs, the peasants demanded abolition of the segregated school system, relaxation of the pass laws, and direct representation in parliament.

The regime's local administration collapsed for a time, while Intaba's authority covered some 1,544.4 square miles, embracing a population of 180,000. As the revolt progressed, closer ties were established with the ANC. Anderson Ganyile, a member of the ANC Youth League, served as Intaba's secretary. The Pondo movement adopted as its program the Freedom Charter, which is the ANC's program as well.

This revolt was crushed only by massive repression, involving massacres, widespread detentions, a state of emergency, and heavy police patrols.

- Tembuland, in the western Transkei, which flared up in sporadic popular actions in 1962. These protests, provoked by land "rehabilitation" schemes, lasted into the following year.

Impact on ANC

These rural revolts played a key role in shaping the political evolution of the main liberation organization, the ANC. Up through the end of the 1940s, the central ANC leadership generally avoided mass action and paid only sporadic attention to the problems and struggles of the countryside. The ANC's political influence in the reserves was thus very limited and in some cases was superseded by that of other political currents.

But under the impact of the growing opposition to the National Party's apartheid policies, the ANC was transformed. A new generation of younger leaders emerged, organized at first within the ANC Youth League. They pressed for the ANC to adopt a perspective based on mass mobilization and uncompromising opposition to all racist policies. Several of them, including Nelson Mandela, Walter Sisulu, and Govan Mbeki, came from the Transkei and were thus more aware of and sensitive to the plight of the rural masses. A manifesto of the Youth League in 1948 called for "far-reaching agrarian reforms," including a "re-division of land among farmers and peasants of all nationalities in proportion to their numbers."

By the end of the 1940s, these younger activists had come into the national leadership of the ANC itself. They helped reorient it politically. The ANC's new course included a much greater reliance on mass action, in place of the earlier leadership's emphasis on negotiations with the authorities over hoped-for reforms; recognition of the value to the African majority of forging alliances with the other sectors of the oppressed Black population (Indians and Coloureds), as well as with anti-apartheid whites; and an increased interest in and contact with the struggles of the countryside.

In 1952, the ANC launched a national campaign of mass demonstrations, strikes, and civil disobedience known as the Defiance Campaign, organized under Mandela's direction. Of the six laws specifically targeted by the campaign, two directly related to the grievances of the rural masses—the Bantu Authorities Act and the Stock Limitation Proclamation, which empowered the regime to forcibly reduce the number of cattle in the reserves.

Thousands of volunteers symbolically broke the six target laws and presented themselves for arrest. Some African farm workers took part in the campaign in the eastern Transvaal, but the greatest rural response was in the Peddie region of the Ciskei, which saw mass resistance to the regime's cattle-control measures.

James Njongwe, an ANC leader in Port Elizabeth, told ANC supporters during the Defiance Campaign: "Your duty is now to go and spread the message of freedom to the people in the reserves. They know what oppression is, what it is to have their cattle killed. They know what has been done to their chiefs and they are ready. They have been ready for years waiting for you."

During the course of the campaign, the ANC's national membership skyrocketed, from 7,000 to 100,000 dues-paying members. Many of the new ANC branches, particularly in the Cape, were in rural areas.

Freedom Charter

Over the next few years, the ANC helped lead political discussions on the program for South Africa's national, democratic revolution. In preparation for a Congress of the People, ANC and other political activists approached a wide variety of individuals and organizations, both urban and rural, for their demands, proposals, and suggestions.

"We call the farmers of the reserves and trust lands," declared a call for the congress. "Let us speak of the wide land, and the narrow strips on which we toil. Let us speak of brothers without land, and of children without schooling. Let us speak of taxes and of cattle, and of famine. Let us speak of freedom."

When the Congress of the People convened on June 25, 1955, in Kliptown, near Johannesburg, nearly 3,000 delegates attended from around the country. They adopted a program for a democratic South Africa, called the Freedom Charter. The section of the charter entitled, "The land shall be shared among those who work it!" declared:

> Restrictions of land ownership on a racial basis shall be ended, and all the land redivided amongst those who work it, to banish famine and land hunger;
> The state shall help the peasants with implements, seed, tractors, and dams to save the soil and assist the tillers;
> Freedom of movement shall be guaranteed to all who work on the land;
> All shall have the right to occupy land wherever they choose;
> People shall not be robbed of their cattle, and forced labor and farm prisons shall be abolished.[6]

The Freedom Charter's land policy—that everyone who actually tills the soil is entitled to land—was consistent with its overall nonracial democratic stance, summed up by the slogan that "South Africa belongs to all who live in it, Black and white."

Although the ANC as a whole adopted the Freedom Charter the following year, some within it, styling themselves "Africanists," rejected this perspective. They soon split from

---

6. For the full text of the Freedom Charter see "The Freedom Charter," *New International* no. 5, pp. 99–105 (2014 printing).

the ANC to form the Pan Africanist Congress (PAC). Potlako Leballo, a key PAC leader, stated in late 1957 that the Africanists favored "the restoration of the land to its rightful owners—the Africans." This excluded not only white working farmers, but also Coloureds and Indians. The ANC, on the other hand, insists that it will guarantee the rights of white working farmers to keep working their land and provide an equal opportunity to farm to all South Africans who choose to do so, on a nonracial basis.

The ANC's incorporation of the struggles of rural Blacks into its overall political perspective was soon reflected in increased ANC activity and influence in the countryside. Although little ANC involvement was apparent in the peasant revolts of the 1940s and early 1950s, by the end of the decade it was increasing. But this was a process, and the ANC's direct role was at times still limited.

According to Govan Mbeki, it was not until the 1960 Pondo revolt in the Transkei that the weight of the struggle in the countryside was fully appreciated by the ANC leadership as a whole. "The Pondo movement," Mbeki wrote, "succeeded by example in accomplishing what discussion had failed to do in a generation—convincing the leadership of the importance of the peasants in the reserves to the entire national struggle."[7]

Organizing farm workers

Along with this increased attention to rural struggles came the first serious effort to unionize African farm laborers.

The groundwork was laid during the 1950s by Gert

---

7. Govan Mbeki, South Africa: *The Peasants' Revolt* (Harmondsworth: Penguin Books, 1964), p. 130.

Sibande, president of the Transvaal ANC. Banned by the government from living in urban areas, he turned toward building up the ANC in the Transvaal's countryside. Specific complaints by farm workers were passed on to the South African Congress of Trade Unions (SACTU), a nonracial but largely Black union federation that was part of the ANC-led Congress Alliance.

In 1960, SACTU formed an agricultural workers' organizing committee in the Transvaal. Besides Sibande, it included Elijah Mampuru, a former leader of a peasants' organization called Sebatakgomo (a call to arms in the Sotho language).

The following year, in October, the Farm, Plantation, and Allied Workers Union (FPAWU) was launched. It later affiliated to SACTU. The delegates at the FPAWU's inaugural congress projected a struggle for significantly higher agricultural wages, which at that time averaged £2 to £3 *a month* for African farm workers. The FPAWU issued a demand for £1 a day for those agricultural workers who earned only cash wages. For workers who received both wages and land, the demand was for £5 a month, plus more than ten acres of fertile land, with farming implements provided by the white farm owner.

Recruitment to the FPAWU was slow because of the difficulties of organizing farm laborers—who were scattered in isolated farms and regions throughout the countryside. But there was steady growth, nonetheless. The Food and Canning Workers Union, another SACTU affiliate, also aided farm workers in the Western Cape.

Repression and resurgence

Confronted by this rural unrest—on top of the massive anti-apartheid mobilizations in the cities and towns—the regime clamped down even harder.

Following the Sharpeville massacre of March 1960, the apartheid authorities outlawed the ANC, PAC, and other political groups. The ANC launched an armed struggle, but its initial guerrilla operations were crushed. By 1963 top ANC leaders such as Mandela, Sisulu, and Mbeki had been arrested. They were later sentenced to life in prison, where they remain today. Others, such as Oliver Tambo, had already gone into exile to avoid arrest.

Although SACTU was not formally outlawed, it was crippled by the repression. Nothing more was heard of the FPAWU after 1964.

Stepped-up repression in the Bantustans took a heavy toll. Thousands were jailed and dozens executed. In the Transkei, the state of emergency first imposed in 1960 was extended through most of the next two decades. In 1963, candidates of the Eastern Pondoland People's Party, led by surviving activists from the Pondo revolt, were arrested and the party was crushed; the authorities accused it of being in contact with the exiled ANC leadership.

The apartheid regime succeeded by the mid-1960s in crushing almost all overt opposition in the country. That facilitated its efforts to build up the subservient Bantustan administrations and carry through yet more massive forced resettlements.

Compelled to function largely outside the country for the time being, the ANC concentrated on training the fighters of its military wing, Umkhonto we Sizwe (Spear of the Nation), and on strengthening its international and diplomatic work.

It also further elaborated its program for the South African revolution, within the framework set by the Freedom Charter. At a conference held in 1969 in Morogoro, Tanzania, the ANC discussed and adopted a series of documents on program, strategy, and tactics. Among them was a more detailed explanation of the various clauses of the Freedom Charter, including the charter's agrarian program (see appendix).

By the early 1970s, both urban and rural Blacks within South Africa were beginning to recover from the repression of the previous decade. Although the ANC remained banned, there arose new Black political groups that were able to function openly, primarily among students and urban township residents.

Some Bantustan officials hinted at the discontent that was also simmering just below the surface in the Bantustans. Kaiser Matanzima, appointed by Pretoria to head up the Transkei, warned in 1973, "Among the youths there is a movement towards Black Power. I advise the whites to reason with us. We will not grab the land we want—but our youths will take the land by force."

Spurred by the victory of Cuban volunteer forces and Angolan freedom fighters over invading South African troops in early 1976, massive youth rebellions swept South Africa later that year. The largest and most prominent actions took place in Soweto and other urban townships, where hundreds of thousands poured into the streets. But the unrest quickly spread to some of the Bantustans, such as BophuthaTswana, Venda, Lebowa, and QwaQwa, involving mainly students and other residents of townships located within their borders. In addition, several dozen white-owned farms in the Transvaal, Natal, Orange Free State, and Cape were set to the torch. Among them was the farm of Hendrik Schoeman, Pretoria's minister of agriculture, which suffered more than $100,000 in damages.

Although the 1976 rebellions were eventually suppressed—at a cost of more than 600 lives—they opened a new period in South Africa, in which open, mass anti-apartheid activity has again become possible. New political and community groups have been formed, and the Black trade union movement has expanded rapidly. Blacks throughout the country, including in rural areas, have been inspired to advance their own particular struggles.

'Why should we move?'

One of the most prominent forms of rural resistance has been the opposition by Black farming communities to forcible resettlement. Throughout the "white" countryside there remain several hundred pockets of Black-owned land, officially called "Black spots." The regime's policy is to eliminate them.

Although little is known of most of these struggles—given their relative isolation from the main urban centers—some of the more tenacious ones have achieved prominence.

At Mgwali, in the Eastern Cape, 4,000 of the 5,000 residents actively oppose the regime's efforts to deport them to the nearby Ciskei—where they would lose most of their land rights. They have formed the Mgwali Residents Association to fight the eviction and have held rallies and organized petition campaigns. They have had to confront not only Pretoria's police but also those of the Ciskei administration.

In KwaNgema, in the eastern Transvaal, another 5,000 people are fighting removal from land their community has occupied for more than 125 years. The regime claims they must leave to make way for a dam. The residents have

Peasant women protest against pass laws, western Transvaal, 1957.

elected a community council that is strongly opposed to the move.

Near Bergville, in Natal Province, the regime also used a dam construction project as a pretext for removing farming communities from three "Black spots" and part of KwaZulu. But their resistance forced the authorities to buy them new farms in the same area.

One of the most concerted struggles has taken place at Driefontein, in the southeastern Transvaal. It is home to more than 5,000 Africans living on land originally bought in 1912 by Pixley ka Isaka Seme, a founder of the ANC. Although many Driefontein residents must work outside the area as migrants, others are able to carry out some farming activities.

Among the reasons the government has given for trying to remove the Driefontein community are that it is "badly situated" and that nearby white farmers are angry that Driefontein competes with them for farm labor.

The residents of Driefontein have resisted resettlement since mid-1981. In a petition to the government, they declared, "We, the undersigned landowners and tenants at Driefontein, protest most strongly against the Government attitudes. . . . We and our families have lived at Driefontein for over 70 years. We cannot accept that the Government can simply take our land without being prepared to discuss it with us."

Saul Mkhize, chairperson of the Driefontein residents' council, told a foreign journalist, "We paid for our land and we wish to keep it. We will not own the new land to which we are supposed to move. We will merely be squatters, and who knows when someone else will decide to move us again? Why should we move? Because the government wants our land for their own purposes? For the minerals beneath the ground? Would they move white people in this way—by buses to barren land with no roads, no water, no schools, no shops, nothing?"

In March and April 1983, large meetings were held to protest the resettlement moves. At the second of these, Mkhize was shot dead at point-blank range by a white policeman.

Rather than intimidating the Driefontein residents, Mkhize's murder heightened their resolve. It also brought them considerable solidarity. About 4,000 people turned out for Mkhize's funeral, including supporters of the Congress of South African Students, the Soweto Committee of Ten, and other groups. Bishop Desmond Tutu conducted the ceremony.

General solidarity with communities fighting forced resettlement grew considerably in the wake of Mkhize's murder. The Black Sash, a white women's group opposed to apartheid policy, organized a protest vigil in Johannesburg to focus attention on forced resettlements. The Association for Rural Advancement has assisted communities fighting eviction in Natal Province. Alan Boesak, president of the World Alliance of Reformed Churches and a leader of the United Democratic Front, spoke at a November 1983 rally in Mogopa, another community fighting to keep its land. The Progressive Federal Party, the main white bourgeois opposition party, raised a motion in the Transvaal Provincial Council calling for a halt to all forced resettlements.

Even the generally pro-apartheid Natal Agricultural Union, a white farmers' group, has questioned the regime's drive against the "Black spots." Its president stated in September 1984, "To leave the areas alone and to invest money to improve them would benefit the local people who would not have to move and would help to ensure better job opportunities."

Under such national pressure, the apartheid regime backed down. In January 1985 Minister of Cooperation and Development Gerrit Viljoen promised that the resettlement program would be "suspended."

But those communities that had been threatened with resettlement have not let down their guard. In early February, just a week after Viljoen's announcement, representatives of ninety-five such communities in the Transvaal, northern Cape, and Orange Free State met near Johannesburg. "It is not enough that the government says it will reconsider some of the areas," the conference resolved. "Ourselves as well as other communities threatened have the right to stay where we are. We will fight for our future whatever reprieves and threats the government issues. We believe that it is our struggle which has shown the government that to continue with removals will cause bloodshed."

Protest Bantustan 'independence'

In the Bantustans themselves, political and social unrest has become much more frequent and widespread since the late 1970s.

In part, this has been spurred by the regime's drive to proclaim a number of the Bantustans "independent" states, as has happened to the Transkei, Ciskei, BophuthaTswana, and Venda. This "independence" is a fraud, since these fragmented areas are integral parts of South Africa and their administrations are politically subservient to Pretoria. Throughout the country Blacks have opposed this "independence," viewing it as a smoke screen for denying them their full citizenship rights in South Africa as a whole and as a maneuver to keep Blacks divided among themselves along language and tribal lines. They have also seen it as a further step toward entrenching the unpopular Bantustan officials.

All the Bantustans, "independent" or not, have felt the impact of the massive anti-apartheid upsurge rocking the country. This has been especially true of those Bantustans that have large urban concentrations.

The Ciskei authorities have for several years been engaged in a bitter conflict with the rebellious residents of Mdantsane, a large Black township located within its borders, a few miles from the industrial city of East London. In late 1983, when Mdantsane residents launched a bus boycott to protest high fares, the administration of Lennox Sebe unleashed its police, who killed as many as ninety people. Independent Black trade unions have a significant base in Mdantsane, especially the South African Allied Workers Union (SAAWU). Sebe banned the SAAWU, accusing it of being a front for the ANC. Since then, scores of trade unionists have been arrested in the Ciskei.

Mass demonstrations, marches, student rallies, consumer boycotts, and workers' actions have also shaken Black townships in BophuthaTswana (near Pretoria) and in KwaZulu (near Durban).

Increasingly, however, such resistance has been spreading into the more rural parts of the Bantustans as well.

Transkei 'restive'

Although the Transkei has been the scene of some of the most brutal repression in the country, it continues to experience persistent political and social ferment.

In the late 1970s, Transkei officials were openly complaining about opposition among rural residents, including some chiefs, to plans to turn over more land to government-run tea plantations. New livestock levies, designed to force cattle herders to sell some of their cattle on the market, were widely resisted.

Opposition to Matanzima's Transkei administration among the Tembu people has been heightened by Matanzima's long-standing conflict with Tembu Paramount Chief Sabata Dalindyebo.

Since the early 1960s, Dalindyebo has sharply opposed aspects of Pretoria's apartheid policies in the Transkei, such as its land "rehabilitation" schemes. He has also condemned the regime's policy of bestowing chiefly titles on its collaborators (such as Matanzima), a policy, he maintained, that "perverted and prostituted chieftainship." When Matanzima decided to accept "independence" for the Transkei in 1976, Dalindyebo opposed that as well. In 1979 he united several opposition forces within the Transkei to form the Democratic Progressive Party (DPP).

That same year Dalindyebo was arrested and charged with slandering the Transkei administration. Demonstrations of up to 5,000 protested Dalindyebo's arrest, and leaflets were distributed in Umtata, the Transkei's capital, denouncing the sham of "independence." Under this pressure, Matanzima was forced to let Dalindyebo off with a fine. But he kept up the pressure, arresting many DPP activists. Dalindyebo himself was stripped of his position as the Tembu paramount chief. Again facing imminent arrest in late 1980, he chose to flee into exile instead.

On December 3, 1980, Dalindyebo appeared at a news conference in Lusaka, Zambia, with ANC President Oliver

Tambo at his side. He proclaimed his alignment with the ANC, noting that his grandfather had been a founder of the ANC and that Nelson Mandela was a cousin of his.

"Throughout the Transkei the people are restive," Dalindyebo said. "They want to do something to bring their suffering to an end." He urged the people of the Transkei to "continue the struggle, refuse to submit to the terrorism of the Matanzima brothers. The struggle has to be conducted in the Transkei, and knowing the people of this area, I am convinced that they will stand on their feet in their millions and, together with the rest of the people of [South Africa], sweep away the Matanzimas and their bosses in Pretoria."

Pondoland, the site of the massive peasant rebellion of 1960, has also remained a source of opposition to the Matanzima administration. Through the late 1970s, there were sporadic armed clashes and attacks on police. Hundreds were arrested. In March 1984, Transkeian police attacked a mass gathering in East Pondoland, killing three and arresting dozens on charges of holding an illegal meeting. This provoked fierce protests. In December of that year, the people of Tsolo district, also in Pondoland, rebelled. Six people were killed in the fighting, including a tribal chief, and more than 200 were later arrested. By early 1985 protesters were frequently erecting roadblocks on the main road through Umzimkulu, another part of the Transkei.

'They will be prepared to take up arms'

The opposition in KwaZulu to apartheid's Bantustan policy is also beginning to break out into occasional clashes in some rural areas.

This has been provoked to an extent by Pretoria's Bantustan "consolidation" program, in which hundreds of thousands of Africans are to be moved as KwaZulu's borders are redrawn to make it a more cohesive geographical entity. In 1981, Oscar Dhlomo, a KwaZulu cabinet member, warned, "Already some people are saying they would go into the bush rather than move. That means they will be prepared to take up arms to protect their land."

Cattle thefts from nearby white-owned farms have increased sharply, reflecting both the desperate poverty of the KwaZulu residents and their desire to recover herds that had been taken from them by the white conquerors. "We think that the white government should know that these vast herds of cattle on the white farms adjoining us will be used to feed us pretty soon," one tribal *induna* (headman) from a drought-stricken area of KwaZulu openly declared at a news conference in 1981. "Nobody will be prepared to die while there is food next door."

White capitalist farmers bordering on KwaZulu have complained that they have lost a large portion of their income through livestock thefts. Compounds for farm laborers have also been burned down, and there have been growing reports of armed Africans sighted in white farming areas. A particularly racist white farmer was assassinated in 1983; he had previously killed a KwaZulu resident who strayed onto his land.

The KwaZulu administration, headed by Chief Gatsha Buthelezi, announced plans in 1983 to introduce new repressive measures against people who attended or promoted armed assemblies.

Although Buthelezi is bitterly opposed to the UDF and ANC, he has nevertheless come into conflict with Pretoria's policies from time to time. One such occasion came in 1982 when the regime announced plans to arbitrarily cede the Imgwavuma area of KwaZulu to Swaziland, an independent country between South Africa and Mozambique. This aroused considerable opposition among Blacks throughout South Africa, including in Imgwavuma and the rest of KwaZulu. When government minister Piet Koornhof arrived in Ulundi, the KwaZulu capital, to explain the move, he was met by thousands of angry protesters. The regime later backed down and decided to abandon its plan.

The Bantustan of Lebowa, in the northeastern Transvaal, has seen a revival of resistance by members of the Matlala tribe, who have been fighting against efforts to restrict their land and grazing rights for more than thirty years. They are commonly known as the "Congress people," and some are former members of the ANC.

Their revolt, which began in the early 1950s, was originally sparked by government attempts to impose an appointed chief, following the adoption of the 1951 Bantu Authorities Act. They also opposed land "planning" programs, which have been used as pretexts for decreasing the size of farming plots and cattle herds. They refused to pay taxes, bring their cattle in for inoculation, or move into the areas the government had demarcated for them. After three years of resistance, they won, and the authorities left them alone for the next twenty-five years.

But in 1978 the government began a new drive to try to bring the "Congress people" under its control and subject them to its agricultural "planning" regulations. Once again they resisted. The home of a chief whom they accused of "collaborating with the Lebowa government" was stoned and burned in late 1979. Early the next year the authorities responded with mass arrests, beatings, house burnings, and cattle confiscations.

Elsewhere in Lebowa, tens of thousands of members of the Batlokwa tribe have been fighting ever since 1978 against government efforts to move them to another location.

Worker-peasant alliance

This ferment in the countryside is an integral part of the broader struggle of the oppressed Black majority against apartheid rule. As it has advanced, it has won greater solidarity in the urban centers and has encouraged various political and union groups to take up and champion the demands of the rural masses.

The demand of the ANC that South Africa's vast lands be opened up to anyone who wants to farm is in the interests of not only all oppressed Black South Africans. It is also in the interests of white workers and working farmers farsighted enough to understand the better life that will be made possible by the destruction of apartheid and its replacement by a nonracial democratic society, a society free of segregation and discrimination, in which all will enjoy equal rights.

All Blacks have a stake in fighting to put an end to laws that restrict African farming rights. Even the National African Federated Chamber of Commerce, a politically conservative association of African businessmen, has criticized these racist restrictions.

But this issue is of particular importance to Black workers. They were forcibly proletarianized through the theft of their land and the regime's denial to them of any other source of livelihood. In a democratic South Africa, some may wish to return to the land, to be free and independent commodity producers. A deep-going agrarian reform would thus immeasurably strengthen the worker-peasant alliance that is already being forged in struggle against the apartheid state.

South Africa's 2 million migrant workers form one of the most important direct links between the struggles in the countryside and in the cities. In the Bantustans they and their families are confronted with the same land shortages and poverty that affect all Bantustan residents. In the cities and mines they are subjected to the same unfree labor conditions and superexploitation that afflict all Black workers.

The employers have used the existence of the Bantustans to try to prevent or hinder unionization among migrant workers. Strikers and union organizers have often been fired and deported back to the Bantustans, while the bosses have also sought to recruit scab labor from among the millions of Bantustan unemployed.

Despite these difficulties, migrant workers have begun to organize unions. During a two-day general strike in the Transvaal in November 1984 to protest police repression, unionized migrant workers participated to as great an extent as nonmigrant workers did. The National Union of Mineworkers (NUM)—which is composed mostly of migrant workers—is today one of the most powerful Black unions in the country. In addition to other issues, it has fought the efforts of the bosses to extend the institutions and divisions of the Bantustans into the mine compounds themselves.

Agricultural laborers are also a key link. Many are themselves part worker and part peasant, laboring for a white farm owner for part of the year in return for permission to farm a plot of white-owned land the rest of the time.

Since the suppression of the SACTU-affiliated Farm, Plantation, and Allied Workers Union in the 1960s, no new agricultural workers' unions have yet been able to emerge. But some farm workers have been drawn into struggle recently, in part through the efforts of nonagricultural unions. In 1980, the Food and Canning Workers Union supported a strike by 1,000 workers (most of them women) at an apple cooperative in the Western Cape and negotiated on their behalf. When the NUM launched a boycott of local mine concession stores in East Driefontein in February 1985, it won the support of local farm workers.

A number of the predominantly Black nonracial trade unions support the perspective of the Freedom Charter, including its program for the land.

At its founding conference in Durban in late 1985, the Congress of South African Trade Unions (COSATU)—today the biggest union federation in South Africa—adopted a resolution rejecting the entire Bantustan system and favoring the creation of "a democratic and unitary South Africa." It also placed a priority on organizing agricultural laborers.

The United Democratic Front has likewise come out strongly against the Bantustans. Speaking at a June 1984 rally in Johannesburg to protest forced population removals, Rev. Frank Chikane, the UDF vice-president, called the Bantustan policy "an evil, unjust, satanic policy." He declared that the "Bantustan leaders give credibility to a sin. They are party to the crime of dispossession. They are guilty of the pain suffered by millions of people who were removed. They are guilty of the division of people—urban against rural, Zulu against Sotho, black against white."

In April 1985, the UDF General Council issued a platform of demands, the first point of which was: "The immediate scrapping of the 1913 and 1936 Land Acts and all Group Areas laws, and an end to any form of forced removals." Though most of the 600 groups affiliated to the UDF are based in the cities and towns, there are some rural affiliates as well. Regional UDF structures have been set up in the predominantly rural northern Cape and Orange Free State.

In late 1984, several thousand people attended a "Rural People's Rally" in the northern Transvaal, many of its participants traveling long distances from different Bantustans to attend. Another protest, organized by the UDF Northern Transvaal Area Committee, was held in April 1985 in Pietersburg, a town in the heart of a white farming region. Demonstrators carried placards denouncing the Bantustans, as well as a police massacre in Uitenhage a month earlier. In November, a mass boycott was launched against white-owned shops in Pietersburg to protest the lack of adequate irrigation and grazing fields in nearby Lebowa and to demand the lifting of the state of emergency and an end to the police repression.

ANC's growing rural support

The importance of the struggle in the countryside has been emphasized repeatedly by the African National Congress. A May 5, 1985, broadcast into South Africa over the ANC's Radio Freedom declared, "Many rural areas in the Transvaal, Orange Free State and the Cape have in recent days seen the populations there becoming more and more restive.... Frustrations are running high and the people are no longer prepared to be governed in the same old way....

Violent confrontations which have been going on in the urban areas between residents and the regime's armed forces have come to the masses of our people in the rural areas."

The broadcast concluded, "The populations in the rural and urban areas are one. Let there be concrete and active solidarity among the oppressed."

Parallel to the ANC's reemergence as the most popular and influential organization among urban Blacks, there have also been signs of its growing presence in the Bantustans.

Officials of both the Ciskei and Transkei often charge that ANC "terrorists" are active in those Bantustans, and they have jailed political activists accused of being ANC members. In April 1983 Kaiser Matanzima's brother George, who is also a key figure in the Transkei administration, maintained that the opposition Democratic Progressive Party was an ANC "agent," and that the ANC had built up cells in many of the Transkei's villages and towns. Recent surveys at the University of the Transkei have found that a majority of the students openly support the ANC.

In Venda, guerrillas attacked a police station in Sibasa in November 1981, and Venda police clashed with suspected insurgents two years later. A number of local peasants were arrested on charges of assisting the guerrillas.

Several trials are now under way of people arrested in KwaZulu on charges of being ANC guerrillas or of providing them with assistance. One of the accused, Malinga Zondo, had been a high KwaZulu official. In another case, twelve men and a woman are charged with setting up bases in the Ingwavuma region of KwaZulu for the "recruitment and training of the local populace." Four of the accused admitted in court that they were members of Umkhonto

Sugar plantation workers in Natal.

Large sections of South Africa's oppressed Black population are consigned to miserable poverty.

we Sizwe, the ANC's armed wing.

As the countrywide popular mobilizations against the apartheid regime escalated during the course of 1984 and 1985, the ANC placed greater stress on the need to oppose the Bantustan system and make its administration unworkable.

The struggle against the Bantustans, ANC President Oliver Tambo said in a January 8, 1985, speech, is inextricably tied to the "solution to the land question."

"The dispossession of our people of the land that is theirs remains one of the most burning national grievances," he said. "We repeat our call to our people to give serious attention to the organisation and mobilisation of our rural masses. Basing ourselves on the needs of the people, and taking due account of the concrete conditions of their existence, we must devise suitable organisational structures and mechanisms to reach our rural masses and provide them with the organisational and political tools to defend themselves against exploitation and to assert their right to the land."

The perspective that must be placed before the people of the countryside, Tambo said, is that of "seizing the land from the dispossessor."

APPENDIX

The land shall be shared among those who work it!

The indigenous people of South Africa, after a series of resistance wars lasting hundreds of years, were deprived of their land. Today in our country all the land is controlled and used as a monopoly by the white minority. It is often said that 87 per cent of the land is "owned" by the whites and 13 percent by the Africans. In fact the land occupied by Africans and referred to as "reserves" is state land from which they can be removed at any time but which for the time being the fascist government allows them to live on.

The Africans have always maintained their right to the country and the land as a traditional birthright of which they have been robbed. The ANC slogan "Mayibuye i Africa" [Come back, Africa] was and is precisely a demand for the return of the land of Africa to its indigenous inhabitants. At the same time the liberation movement recognizes that other oppressed people deprived of land live in South Africa. The white people who now monopolize the land have made South Africa their home and are historically part of the South African population and as such entitled to land. This made it perfectly correct to demand that the land be shared among those who work it.

This is an excerpt from the African National Congress presentation at the May 1969 Morogoro Conference.

But who works the land? Who are the tillers?

The bulk of the land in our country is in the hands of land barons, absentee landlords, big companies, and state capitalist enterprises. The land must be taken away from exclusively European control and from these groupings and divided among the small farmers, peasants, and landless of all races who do not exploit the labor of others. Farmers will be prevented from holding land in excess of a given area, fixed in accordance with the concrete situation in each locality.

Lands held in communal ownership will be increased so that they can afford a decent livelihood to the people and their ownership shall be guaranteed. Land obtained from land barons and the monopolies shall be distributed to the landless and land-poor peasants. State land shall be used for the benefit of all the people.

Restrictions of land ownership on a racial basis shall be ended, and all land shall be open to ownership and use to all people, irrespective of race.

The state shall help farmers with implements, seeds, tractors, and dams to save soil and assist the tillers. Freedom of movement shall be guaranteed to all who work on the land. Instruments of control such as the "trek pass," private jails on farms, forced labor shall be abolished. The policy of robbing people of their cattle in order to enforce them to seek work in order to pay taxes shall be stopped.

BUILDING A PROLETARIAN PARTY

The Low Point of Labor Resistance Is Behind Us
The Socialist Workers Party Looks Forward

JACK BARNES
MARY-ALICE WATERS
STEVE CLARK

The global order imposed by victors of the inter-imperialist slaughter of World War II is shattering, with explosive ramifications for workers and farmers worldwide. A long retreat by the working class and unions has come to an end. More and more workers of all ages, skin colors, and both sexes are saying, "Enough is enough!" This book highlights opportunities ahead for class-conscious workers to forge a labor party built on the unions. And a mass proletarian vanguard able to lead the struggle to end capitalist rule, opening a future for humanity. $10. Also in Spanish and French.

Labor, Nature, and the Evolution of Humanity
The Long View of History

FREDERICK ENGELS, KARL MARX
GEORGE NOVACK, MARY-ALICE WATERS

Without understanding that social labor, transforming nature, has driven humanity's evolution for millions of years, working people are unable to see beyond the capitalist epoch of class exploitation that warps all human relations, ideas, and values. Only the revolutionary conquest of state power by the working class can open the door to a world free of capitalist exploitation, degradation of nature, subjugation of women, racism, and war. A world built on human solidarity. A socialist world. $12. Also in Spanish and French. Ebook for the blind or low vision: visit Bookshare.org.

WWW.PATHFINDERPRESS.COM

AFRICAN FREEDOM STRUGGLE

Thomas Sankara Speaks
The Burkina Faso Revolution, 1983–87

Under Sankara's guidance, Burkina Faso's revolutionary government led peasants, workers, women, and youth to expand literacy; to sink wells, plant trees, erect housing; to combat women's oppression; to carry out land reform; to join others worldwide to free themselves from the imperialist yoke. $20. Also in French.

Capitalism and the Transformation of Africa
Reports from Equatorial Guinea
MARY-ALICE WATERS, MARTÍN KOPPEL

Describes how, as Equatorial Guinea is pulled into the world market, both a capitalist class and a working class are being born. Also documents the work of volunteer Cuban health-care workers there—an expression of the living example of Cuba's socialist revolution. $10. Also in Spanish and Farsi.

Nelson Mandela Speaks
Forging a Democratic, Nonracial South Africa

Mandela's speeches from 1990 through 1993 recount the course of struggle that put an end to apartheid and opened the fight for a deep-going political, economic, and social transformation in South Africa. $17

How Far We Slaves Have Come!
South Africa and Cuba in Today's World

NELSON MANDELA, FIDEL CASTRO

Speaking together in Cuba in 1991, Mandela and Castro discuss the role of Cuba in the history of Africa and Angola's victory over the invading US-backed South African army. That victory accelerated the fight to bring down the racist apartheid system. $7. Also in Spanish and Farsi.

Malcolm X Talks to Young People

Four talks and an interview given to young people in Ghana, the United Kingdom, and the United States in the last months of Malcolm's life. He discusses imperialist intervention in the Congo and Vietnam, why he stopped using the description "Black nationalism," and more. Concludes with memorial tributes by a young socialist leader to this great revolutionary. $12. Also in Spanish, French, Farsi, and Greek.

The Coming Revolution in South Africa

JACK BARNES

Writing nearly a decade before the fall of the white supremacist regime, Barnes explores the social character and roots of apartheid in South African capitalism and the tasks of toilers in city and countryside in dismantling it as they forge a communist leadership of the working class. Also includes "Why Cuban Volunteers Are in Angola," two speeches by Fidel Castro. In *New International* no. 5. $14. Also in Spanish and French.

WWW.PATHFINDERPRESS.COM

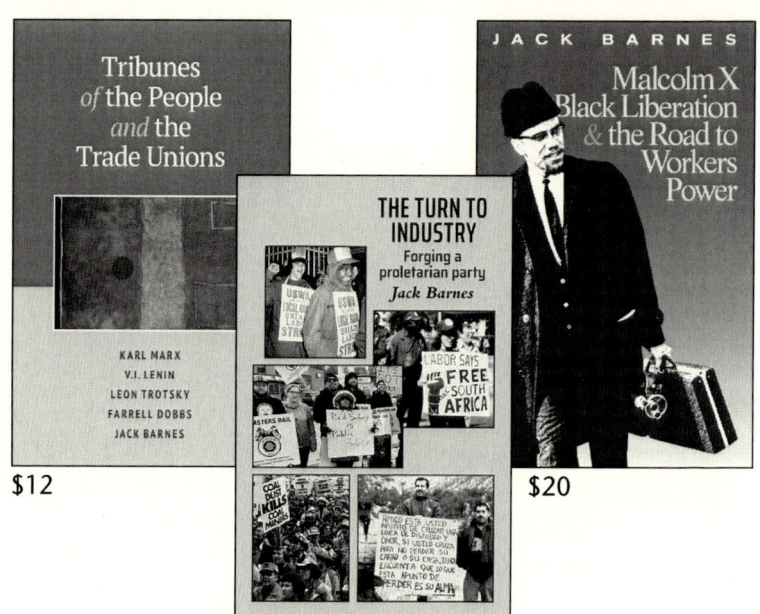

$12

$20

$15

Three books to be read as one ...

about building a party that's working class in program, composition, and action. One that recognizes, in word and deed, the most revolutionary fact of our time ...

... that working people have the power to create a different world as we act together to defend our own class interests—not those of the privileged classes who exploit our labor, not of those who fear us as "deplorables," or just plain "trash."

As we advance along a revolutionary course toward workers power, we will transform ourselves and awaken to our own worth. Also in Spanish and French.

Special Offer!
All three $30

The Turn to Industry and *Tribunes of the People and the Trade Unions* $20

Either book plus *Malcolm X, Black Liberation, and the Road to Workers Power* $25

CUBA'S SOCIALIST REVOLUTION

Women in Cuba: The Making of a Revolution within the Revolution
VILMA ESPÍN, ASELA DE LOS SANTOS YOLANDA FERRER

The integration of women in the ranks and leadership of the Cuban Revolution was intertwined with the proletarian course of the leadership of the revolution from the start. This is the story of that revolution and how it transformed the women and men who made it. $17. Also in Spanish, Farsi, and Greek.

Cuba and the Coming American Revolution
JACK BARNES

This is a book about the struggles of working people in the imperialist heartland, the youth attracted to them, and the example set by the Cuban people that revolution is not only necessary—it can be made. It is about the class struggle in the US, where the revolutionary capacities of workers and farmers are today as utterly discounted by the ruling powers as were those of the Cuban toilers. And just as wrongly. $10. Also in Spanish, French, and Farsi. Ebook for the blind or low vision: visit Bookshare.org.

Colombia: Fidel Castro on the Debate around Revolutionary Strategy and Lessons of the Cuban Revolution
FROM THE PAGES OF THE *MILITANT*

Excerpts from Fidel Castro's *Peace in Colombia* and articles from the *Militant*. In describing the Cuban leadership's efforts to end decades of war between the FARC guerrilla movement and Colombia's brutal regime, Castro in his prologue, afterword, and other statements explains why Cuban revolutionaries, unlike FARC leaders, rejected taking hostages and organized working people to win state power, not pursue a "prolonged people's war." $5. Also in Spanish.

WWW.PATHFINDERPRESS.COM

EXPAND YOUR REVOLUTIONARY LIBRARY

The Jewish Question
A Marxist Interpretation
ABRAM LEON

Why is Jew-hatred still raising its ugly head? What are its class roots—from antiquity through feudalism, to capitalism's rise and current crises? Why is there no solution under capitalism? The author, Abram Leon, was killed in the Nazi gas chambers. Revised translation, new introduction, and 40 pages of illustrations and maps. $17. Also in Spanish and French.

Teamster Rebellion
FARRELL DOBBS

The 1934 strikes that won union recognition for truckers and warehouse workers in Minneapolis and helped pave the way for the working-class social movement that built the industrial unions. The first of four volumes by a central leader of these battles. $16. Also in Spanish, French, Farsi, and Greek.

50 Years of Covert Operations in the US
Washington's Political Police and the American Working Class
LARRY SEIGLE, FARRELL DOBBS
STEVE CLARK

How class-conscious workers have fought against the drive to build the "national security" state essential to maintaining capitalist rule. $10. Also in Spanish and Farsi.

Are They Rich Because They're Smart?
Class, Privilege, and Learning under Capitalism

JACK BARNES

Exposes growing class inequalities in the US and the self-serving rationalizations of well-paid professionals who think their "brilliance" equips them to "regulate" working people, who don't know what's in our own best interest. $10. Also in Spanish, French, Farsi, and Arabic. Ebook for the blind or low vision: visit Bookshare.org.

Lenin's Final Fight
Speeches and Writings, 1922–23

V.I. LENIN

In 1922 and 1923, V.I. Lenin, central leader of the world's first socialist revolution, waged what was to be his last political battle—one that was lost following his death. At stake was whether that revolution, and the international communist movement it led, would remain on the revolutionary proletarian course that brought workers and peasants to power in October 1917. $17. Also in Spanish, Farsi, and Greek.

Is Socialist Revolution in the US Possible?
A Necessary Debate among Working People

MARY-ALICE WATERS

Fighting for a society only working people can create, it is our own capacities we will discover. And along that course we will answer the question posed here with a resounding "Yes." Possible but not inevitable. That depends on us. $7. Also in Spanish, French, and Farsi.

WWW.PATHFINDERPRESS.COM

The Clintons' Anti-Working-Class Record
Why Washington Fears Working People
JACK BARNES

What working people need to know about the profit-driven course of Democrats and Republicans alike over the last three decades. And the political awakening of workers seeking to understand and resist the capitalist rulers' assaults. $10. Also in Spanish, French, Farsi, and Greek.

From the Escambray to the Congo
In the Whirlwind of the Cuban Revolution
VÍCTOR DREKE

Dreke was second in command of the internationalist column in the Congo led in 1965 by Che Guevara. He recounts the creative joy with which working people have defended their revolutionary course—from Cuba's Escambray mountains to Africa and beyond. $15. Also in Spanish.

Cosmetics, Fashions, and the Exploitation of Women
JOSEPH HANSEN
EVELYN REED
MARY-ALICE WATERS

How big business reinforces women's second-class status and uses it to rake in profits. Where does women's oppression come from? How has the entry of millions of women into the workforce strengthened the battle for emancipation, still to be won? $12. Also in Spanish, Farsi, and Greek.

New International

A MAGAZINE OF MARXIST POLITICS AND THEORY

In Defense of Land and Labor

"Capitalist production develops by simultaneously undermining the original sources of all wealth—the soil and the worker." —*Karl Marx, 1867*

THREE ARTICLES

IN *NEW INTERNATIONAL* NO. 13
- **Our Politics Start with the World**
  JACK BARNES
- **Farming, Science, and the Working Classes**
  STEVE CLARK

IN *NEW INTERNATIONAL* NO. 14
- **The Stewardship of Nature Also Falls to the Working Class**
  JACK BARNES, STEVE CLARK, MARY-ALICE WATERS

U.S. Imperialism Has Lost the Cold War
JACK BARNES

The collapse of regimes across Eastern Europe and the USSR claiming to be communist did not mean workers and farmers there had been crushed. In today's sharpening capitalist conflicts and wars, these toilers are joining working people the world over in the class struggle against exploitation. In *New International* no. 11. Also in Spanish, French, Farsi, and Greek.

Capitalism's Long Hot Winter Has Begun
JACK BARNES

In *New International* no. 12. Also in Spanish, French, Farsi, Arabic, and Greek.

$14 each issue

WWW.PATHFINDERPRESS.COM

PATHFINDER AROUND THE WORLD

UNITED STATES
(and Caribbean, Latin America, and East Asia)
> Pathfinder Books, 306 W. 37th St., 13th Floor
> New York, NY 10018

CANADA
> Pathfinder Books, 7107 St. Denis, Suite 204
> Montreal, QC H2S 2S5

UNITED KINGDOM
(and Europe, Africa, Middle East, and South Asia)
> Pathfinder Books, 5 Norman Rd.
> Seven Sisters, London N15 4ND

AUSTRALIA
(and New Zealand, Southeast Asia, and the Pacific)
> Pathfinder Books, Suite 2, First floor, 275 George St.
> Liverpool, Sydney, NSW 2170
> Postal address: P.O. Box 73, Campsie, NSW 2194

JOIN THE PATHFINDER READERS CLUB
BUILD YOUR LIBRARY!

$10 / YEAR
25% DISCOUNT ON ALL PATHFINDER TITLES
30% OFF BOOKS OF THE MONTH
Valid at pathfinderpress.com and local Pathfinder book centers

Go to: www.pathfinderpress.com/products/pathfinder-readers-club